AIR COMBAT PHOTO HISTORY

ATLANTIC AIR WAR:

SUB HUNTERS vs U-BOATS

By
JOHN W. LAMBERT

SPECIALTY PRESS

ISBN: 1-58007-015-9

Library of Congress Catalog Card Number: 98-90888

Text by John W. Lambert

Published by:
Specialty Press Publishers and Wholesalers
11605 Kost Dam Road
North Branch, MN 55056
1-800-895-4585

Book Trade Distribution by:
Voyageur Press
123 North Second Street
Stillwater, MN 55082
Phone: 800-888-9653
Fax: 651-430-2211

Printed in Canada

AIR COMBAT PHOTO HISTORIES

This is the fourth in a series of illustrated World War II histories that will graphically portray the many facets of aerial combat. Some of the rare photos are drawn from government archives and many from private collections of veterans. All show scenes of air action with captions that, as nearly as possible, identify the date, locale, and describe the event or subject. Action pictures were generally made by hand-held cameras and may, understandably, be out of perfect focus. Consider the circumstances.

In every case the photos show real aircraft or combat that occurred between 1940 and 1945, only a short step removed from the technology of World War I. These scenes will not likely be replayed in modern wars with the advent of smart bombs and long range rockets that destroy targets over the horizon, many miles from the point of launch. World War II aerial warfare still had much of the "knights who go forth to do mortal combat" about it. The adversaries were not exactly beating on each other's shields, but it was, in most ways, a terrible sort of "up close and personal" warfare.

The men and aircraft are also depicted in order to give the reader a sense of the nature of the war — brave young men in frail aircraft who went in harm's way. The deeds of warriors on both sides are described here so that future historians may have a record of the violent yet valiant nature of the struggle.

ACKNOWLEDGMENTS

We are indebted to the following individuals and organizations for their assistance in providing photos and information: Ernie Cromie, No. Ireland; Rolf Kolling, Egil Johansen, Lorang Sjonnesen, Norway; Andrew Bird, David Legg, Simon Parry, Air Research Publications and Jerry Scutts from England; Roy J. Hopper, 480th AS Grp; Don Schierenbeck, VB-103; Al Lloyd and Zeke Cormier from the U.S.; the Canadian Armed Forces and Canadian National Archives; Britain's Public Records Office; U.S. National Archives; The Naval Museum, Pensacola, Florida.

TABLE OF CONTENTS

INTRODUCTION

Within the first week of the onset of World War II in Europe German submarines had sunk Allied merchant ships. This particular battle, to strangle England's supply lines, continued with unremitting fury until late 1943 when Allied strength in escort vessels and aircraft, and advances in technology and cryptography turned the tide. Still U-boats continued to sink ships on both sides of the Atlantic until the end of the war in May 1945. The contest for control of the Atlantic was the longest running campaign of World War II.

This vicious battle at sea was prosecuted with scant press coverage and has received comparatively little attention in subsequent history, but it was as critical to the outcome of the war as any major land battle. In Winston Churchill's **The Second World War, Volume 5**, he wrote: "The Battle of the Atlantic was the dominating factor all through the war." As early as September 1939, long before the official U.S. declaration of war, President Franklin Roosevelt, deeply concerned for the survival of England, ordered the U.S. Navy to conduct "Neutrality Patrols," a euphemism for a defense that evolved into an anti-submarine operation.

The convoy struggle was largely fought by the armed trawlers, sloops, corvettes, destroyers, and destroyer escorts of the Allied navies. Early in the conflict aircraft began convoy escort, attempting to probe those voids over the horizon, beyond the escorts, where submarines would congregate. But the first anti-submarine warfare (ASW) patrol planes had serious endurance limitations. Most were land-based bombers designed to attack either strategic industrial targets or to support ground forces. No one had ever conceived an aircraft for the specific ASW role. Royal Air Force (RAF) Coastal Command, therefore, employed virtually everything that could fly in the early stages of the war.

As America entered the Battle of the Atlantic in 1942 its initial contribution was made up of Navy patrol squadrons. Together with the Army Air Force, virtually every type of aircraft, including Navy airships and private aircraft (the Civil Air Patrol), were eventually rallied for convoy escort and U-boat patrols.

By 1943 bombing squadrons equipped with Martin PBM Mariners, Consolidated Catalina PBY flying boats, and the Navy version of the B-24 Liberator (a PB4Y-1 in Navy terminology) and Lockheed PV-1 Venturas had been stationed in the Caribbean, Newfoundland, Iceland, England, and North Africa.

Still there were vast reaches of the Atlantic that could not be patrolled by aircraft from any land base. Beginning in March 1943 a growing fleet of escort carriers and air crews, trained for the sole mission of hunting U-Boats, were deployed in the Battle of the Atlantic. These units operated Grumman TBF Avenger bombers and Grumman F4F Wildcats in teams that tracked down German submarines and struck faster than any surface escort might. By 1944 there were nine U.S. Navy escort carriers sailing in the Atlantic. They proved enormously successful, sinking or joining with surface forces in the destruction or capture of some fifty-six Axis submarines.

Coastal Command ultimately fielded more than a score of ASW units including Norwegians, Czechs, Poles, and French. Britain's Fleet Air Arm provided several units both from shore stations and escort aircraft carriers. Commonwealth squadrons from Canada and Australia were also prominent in the anti-sub campaign, and by 1943 Brazil was patrolling the South Atlantic.

By its very nature sub hunting was low level work. It was rare that submarines could be found on the surface and, when they sighted an attacking aircraft, the U-boat's best defense was to crash dive. Depth charges (bombs that detonated underwater) had, of necessity, to be dropped close by the target or the water absorbed and cushioned the explosion. Hence, the tactic employed by aircraft of all types was diving down and laying the bombs or depth charges as close as possible to the sub or its receding wake.

To counter Allied aircraft, German submarines were later equipped with improved multiple light anti-aircraft cannons. Submarine skippers, who had previously submerged at first sight of a surface warship, would often choose to fight it out with an approaching aircraft. Since the slow moving submarine was a more stable gun platform, the engagements were never one-sided in these unique duels.

At the onset of the war U-boats sailed from German ports. As the Whermacht marched into Europe, additional bases became available in Norway and France. The French ports, Brest, Lorient, St. Nazaire, La Pallice, and Bordeaux had advanced the departure line for the U-boats several hundred miles and eliminated the need for those so based to run the gauntlet between Iceland and the British Isles. French submarine pens, covered with several feet of reinforced concrete, were bombed repeatedly by RAF Bomber Command and the U.S. Army Eighth Air Force with minimal effect. This situation created a necessity for Allied ASW aircraft to attempt placing an aerial umbrella over the Bay of Biscay, a choke point for U-boats transiting to and from French bases.

As Allied attacks in the Bay of Biscay increased, the Germans responded with the employment of Luftwaffe units to assist the submarines by high altitude reconnaissance, and an effort to counter Allied aircraft by direct interception. Here in the Bay of Biscay, some of the most bizarre aerial dogfights of World War II were fought between multi-engine bombers.

The combination of carrier-borne and land-based aircraft and their advanced radar detection, working in close cooperation with surface forces, had won the Battle of the Atlantic by the end of 1944. Germany had lost her French bases a few months after the D-Day invasion and remaining U-boat operations had to be conducted from ports in Norway and Germany. Although the submarines continued to sink ships through 1945, they were never again a threat to Allied victory.

On the following pages we will attempt to present a graphic pictorial history of the intrepid airmen and the aircraft engaged in that long and grim Atlantic war.

GLOSSARY

AAF	U.S. Army Air Force
Asdic	The British term for underwater acoustic detection, sonar in the USN
ASV	Air to Surface Vessel, British term for airborne radar
ASW	Anti Submarine Warfare
BdU	(Befehlshaber der Unterseebotte) Headquarters for Admiral Karl Doenitz's U-boat Command
CAP	Civil Air Patrol
CC	Coastal Command
Cdr.	Abbreviation for rank of Commander
CVE	U.S. Navy term for escort or auxilliary aircraft carrier
Depth Charge	A bomb dropped by surface ships or aircraft that would explode at a preset depth
DD	U.S. Navy term for Destroyer
DE	U.S. Navy term for Destroyer Escort, a small destroyer specifically designed for antisubmarine work
DFC	Distinguished Flying Cross
Ens.	Abbreviation for rank of Ensign in USN
FAW	Fleet air wing, a U.S. Navy unit composed of several squadrons
Fido	A U.S. acoustic torpedo dropped by aircraft to home in on submarines
Flt.	Abbreviation for rank of Flight, as in Flt. Sgt., Flt. Lt.
F/O	Abbreviation for Flying Officer (RAF)
FS	Fighter squadron
Huff-Duff	A high frequency (HF) direction finding (DF) device, or Huff-Duff
(jg)	Abbreviation for junior grade rank, as in Lt. (jg)
KG	Kampfgeschwader, a German fighter unit
Leigh Light	An 80 million candle-power light developed by the British and fitted to both Coastal Command and U.S. aircraft for illuminating a sub at night
Lt.Cdr.	Abbreviation for rank of Lieutenant Commander
Lt.	Abbreviation for rank of Lt. in AAF or USN
MAD	Magnetic Anomaly Detector, an airborne detection device that caught variations in the earth's magnetic field created by a submarine
OTU	Operational training unit, an RAF air unit
PC	173-foot U.S. Navy subchaser
P/O	Abbreviation for rank of Pilot Officer (RAF)
RAAF	Royal Australian Air Force
RAF	Royal Air Force
RCAF	Royal Canadian Air Force
RDF	Radio Direction Finding, an early term for radar
RN	Royal Navy
SC	110-foot U.S. Navy subchaser
SN	Serial number
Snorkel	A device on late model U-boats that allowed air circulation and exhaust discharge while diesels were run below the surface
Sonobuoy	A hydrophone on a float dropped by aircraft, transmitter relayed sounds to the parent aircraft
S/L	Abbreviation for rank of Squadron Leader (RAF)
USCG	United States Coast Guard
USN	United States Navy
Wg	Abbreviation for wing
YAGI	Early radar receiver
ZG	A German squadron of destroyer aircraft

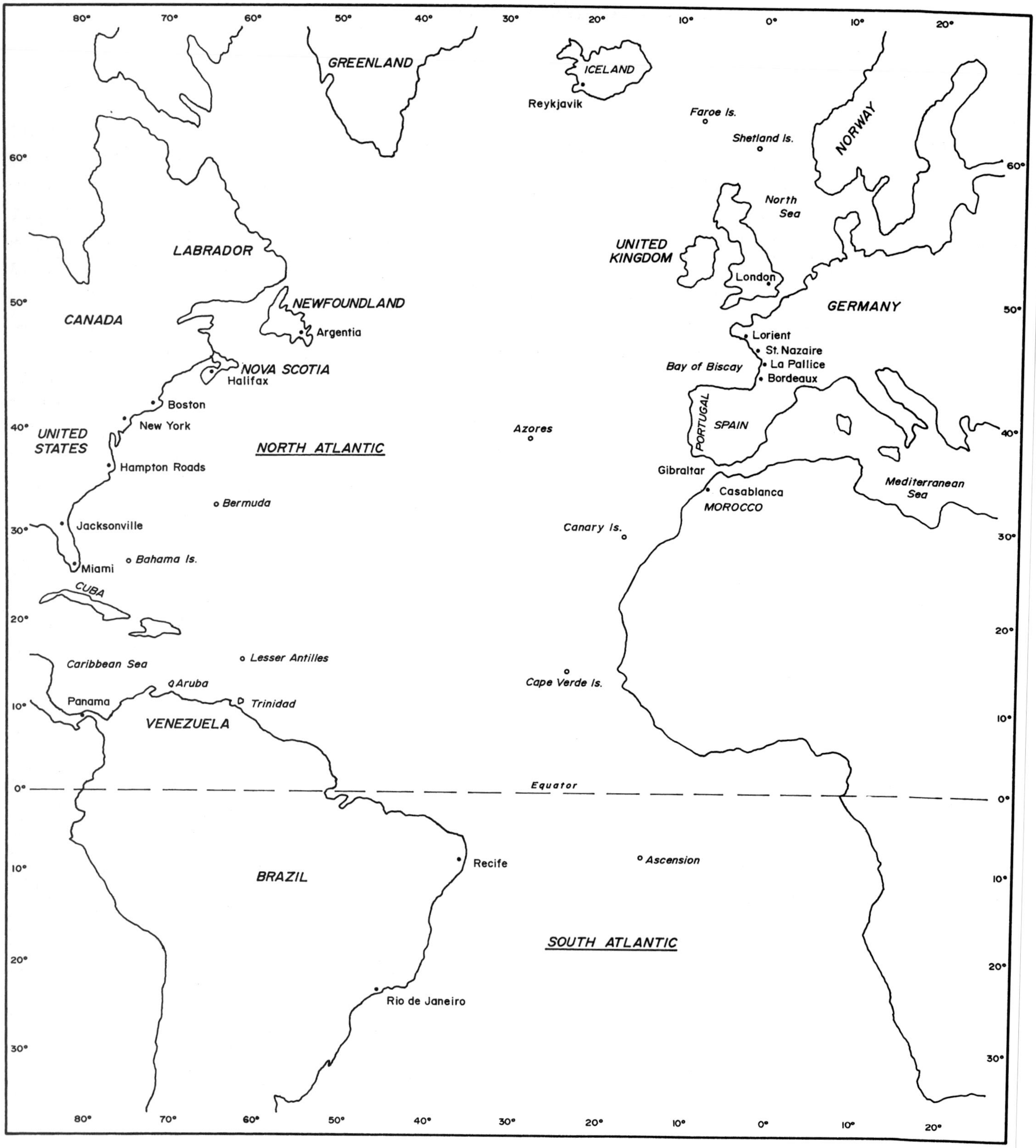
GREENLAND
ICELAND
Reykjavik
Faroe Is.
Shetland Is.
NORWAY
North Sea
UNITED KINGDOM
London
GERMANY
LABRADOR
CANADA
NEWFOUNDLAND
Argentia
NOVA SCOTIA
Halifax
Boston
New York
UNITED STATES
Hampton Roads
NORTH ATLANTIC
Bermuda
Jacksonville
Miami
Bahama Is.
CUBA
Azores
Lorient
St. Nazaire
La Pallice
Bordeaux
Bay of Biscay
PORTUGAL
SPAIN
Gibraltar
Casablanca
MOROCCO
Mediterranean Sea
Canary Is.
Caribbean Sea
Lesser Antilles
Aruba
Trinidad
Panama
VENEZUELA
Cape Verde Is.
Equator
BRAZIL
Recife
Ascension
SOUTH ATLANTIC
Rio de Janeiro

CHAPTER 1

ENGLAND'S LIFE LINE

In the opening four months of World War II — September through December 1939 — German U-boats sank 147 ships. The convoy system was quickly developed to gather the merchantmen under the protection of armed escorts. But the escorts were too few and many less than adequate for the tortuous North Atlantic and, despite their gallant efforts, U-boats sank 520 vessels in 1940, often as many as fifty or sixty per month.

Allied aircraft were of little help in defending the cargo ships from attack; first because there were no aircraft specifically designed for the task and secondly, because those available early in the war had such short range as to limit their worth.

Avro Ansons, soon reduced to training and transport duties, and Handley-Page Hampdens were cast into the never intended anti-sub business. Some, like the Armstrong Whitworth Whitley, had been failures as heavy bombers but had the capacity to haul early radar aloft and with the installation of bomb bay auxiliary fuel tanks, attained considerable range. Others, like the Lockheed Hudson, were variations of prewar transports. All had ranges limited to 300 to 500 miles but could carry a load of depth charges. The main virtue of these early patrol planes being availability near the outset of the war. The great Short Sunderland flying boats, transports dragooned into the antisubmarine war, were the sole exceptions.

Eventually the British aircraft industry modified existing aircraft with ingenious devices for the detection and destruction of submarines and produced long-range bombers for Coastal Command which were role specific ASW aircraft.

Throughout the war England also relied heavily on American production for aircraft that filled the need, primarily Consolidated Catalinas and Liberators and later Grumman Avengers and Martlets.

This No. Atlantic U-boat was one that got away from an Iceland based RCAF Liberator of 10 Sq in late Sep '43. But that particular battle for a pair of convoys cost the Germans four submarines. (Canadian Armed Forces)

Avro Ansons equipped many RAF Coastal Command squadrons at the outset of World War II. The lightly armed, fabric covered aircraft, while beloved as a flying machine, was inadequate for ASW operations. This 269 Sq Anson I was destroyed in a take-off accident in Mar 1940, but the crew survived. By 1941 the unit had converted to Hudsons and was stationed in Iceland. (PRO)

A weather worn Catalina I, SN 550, of RAF 210 Sq. shown early in the war. Her unit and aircraft markings were AL-D. As yet, it sports none of the advanced underwing ASV equipment that would become common on sub hunters. Known as "Canso" to Allied aircrews, the Consolidated flying boats were eagerly sought by the British at the outbreak of the war. Catalinas of 210 Sq sank six U-boats. (Jerry Scutts)

This Fokker T. VIII seaplane was one of ten that escaped to England after the fall of the Netherlands. The crews formed 320 Sq in RAF Coastal Command. (Jerry Scutts)

When Norway fell in June 1940, Norwegian Navy AF crews escaped to England in a handful of their aircraft. One was this German made Heinkel 115 a reconnaissance/torpedo seaplane which was employed for a short time in North Sea U-boat patrols. Note the Catalina in the background. (Egil Johansen)

The Lockheed Hudson, a variation on the Lockheed 14 transport, was one of those early wartime aircraft adapted for the hard pressed British, its chief virtue being availability in large numbers. They served as bombers and transports and many were employed with RAF Coastal Command. Lockheed eventually built hundreds for nine Allied air forces. In addition to its top turret guns, the aircraft had two fixed forward firing and three flexible .303 machine guns, light armament compared to later U-boat AA. Above: A pair of 206 Sq Hudsons low over Heligoland Bight in 1940. Below: The Hudson was still serving in 1943. This aircraft was with 233 Sq out of Gibraltar. (Jerry Scutts)

U-570 was damaged on 27 Aug '41 by a depth charge attack delivered by Sq. Ldr. J.H. Thompson, 269 Sq CC, operating Lockheed Hudsons from Iceland. The damage was not fatal to the vessel but led to the discharge of deadly chlorine gas from batteries, forcing the crew topside. The U-boat captain elected to surrender rather than commit his crew to the icy waters of the North Atlantic. Officers from a British trawler can be seen (center right) conferring with the German skipper. U-570 proved to be an intelligence treasure trove. This photo was taken by a Catalina of 209 Sq. (PRO)

Before Germany invaded Norway in April 1940, the Norwegian government had purchased twenty-four Northrop N3PB reconnaissance seaplanes. Norwegian recruits who had escaped their homeland later trained and manned these planes as "C" Flight, 330 Sq RAF. They were stationed in Iceland on convoy escort and antisub patrol in 1941. (See photos on this and the facing page.) The N3PB had a seven hour range and a top speed of just 160 mph. It was equipped with four fixed, wing mounted guns and a pair of flexible machine guns and could carry three 400 lb. depth charges. Each had a crew of three. The redoubtable Norwegians operated their little Northrops until mid 1942, when the surviving thirteen aircraft were replaced by Catalinas. They had flown over 6,300 hours in the terrible Arctic reaches, claiming several sightings and mounting several attacks on U-boats. (Egil Johansen)

Consolidated Catalinas, furnished by the British, were flown by Norway's 330 and 333 Sq operating with RAF Coastal Command from Iceland, Scotland, and the Shetland Islands. In addition to ASW operations these early model Catalinas flew operations deep inside the Arctic Ocean to Spitzbergen and Jan Mayen Island and also inserted agents into Norway. Here QL-R, of 333 Sq (named "Vingtor") is armed with torpedoes for a shipping strike. (Egil Johansen)

This early model Norwegian 333 Sq. Catalina has been modified for a radar dome just aft of the cockpit. Aircraft KK bore the Norwegian pennant on its side and, not surprisingly, the name "Viking." (Egil Johansen)

An RCAF flight sgt. in the cockpit of his Hudson III. (Jerry Scutts)

Depth charges can be seen falling from a Catalina of RCAF 116 Sq during a 1943 training exercise. (Canadian Armed Forces)

Prior to America's entry into World War II the British contracted for Boeing B-17 Flying Fortress aircraft to shore up RAF Bomber Command. These early C and D models proved to have insufficient defensive armament and lacked the armor that the RAF had found necessary over German controlled Europe. Hence, they were relegated to Coastal Command where their great range was an asset. Above: A Fort I (B-17D), SN AN 537 of 220 Sq. based in Northern Ireland, shepherds a convoy. Below: NR-J of the same unit was a Fort II, SN FL459, an E model, with ASV antenna protruding from the nose and underwing. (Jerry Scutts)

The Boeing Flying Fortress, SN FK201 is a Mark IIA (B-17E) which operated from various Northern Ireland bases, Scotland, and the Azores with RAF 220 Sq. (Ulster Avn. Society)

Resembling a flying porcupine, the Liberator AM 923 was the first version of the famed line, and was operated by RAF 120 Sq from Nutts Corner A/D, Northern Ireland. It featured an early airborne radar known as YAGI with vast antenna from nose, wings, and tail and an interesting 20mm gun pod under the nose. (Ulster Avn. Society)

Above: This Liberator was a Mark V and served in 1943 with RAF Coastal Command 120 Sq flying from bases in Northern Ireland. It was later marked as "BZ8." Ireland based Libs were credited with ten U-boat sinkings in 1943. (Ulster Avn. Society)

Left: A British Armstrong Whitworth Whitley, Mark V proved to be more adept at sub hunting than strategic bombardment, for which it was originally designed. A Whitely scored Britain's first U-boat kill. (NARS)

The Fairey Swordfish began service with the British FAA in 1936 and was obsolete by all standards at the start of the war. Yet it soldiered on, becoming one of the renowned aircraft of WW II. Known as the "Stringbag" to her crews, she served from both land and carriers in virtually every over-water role including ASW. The first occasion in which an Allied escort carrier was involved in the destruction of a sub was on 25 Apr '43 when a Swordfish of 811 Sq from HMS *Biter* assisted DD *Pathfinder* in sinking U-203. (NARS)

A Consolidated Catalina I, from RAF 202 Sq, at its mooring off storied Gibraltar in 1943. (NAM)

Outbound U-71 survived a 5 Jun '42 bombing and strafing by a Sunderland II of 10 (RAAF) Sq. 140 miles west of Bordeaux but was so badly damaged it returned to base. The Sunderland, commanded by Flt. Lt. S.R. Wood, later had an inconclusive engagement with a Focke-Wulf 200. The Aussies returned to base at Mount Batten with many holes in their aircraft and two wounded. (NARS)

Right:
AShort Sunderland of RAF 1435 Flight patrolling off the coast of North Africa in 1942. Early in the war these huge four-engine patrol bombers were adapted from the S.23 flying boats of Imperial Airways, and became one of the most effective ASW weapons. (NARS)

Below: The flight deck of an RAF CC Sunderland (unit Unk.) shows the co-pilot using an Aldus lamp to signal a ship. Behind his seat can be seen Very gun cartridges. (Jerry Scutts)

A Whitley of RAF 10 OTU Sq, Flt. Sgt. J.A. Marsden commanding, sank U-665 on 22 Mar '43 west of St. Nazaire. These tail shots show some of the depth charges being released (left) and part of the spread exploding (below). Weather was so foul that the results of the attack could not immediately be assessed. Postwar investigation determined the victor. (NARS)

An Armstrong Whitworth Mark VII Whitley of CC 612 Sq festooned with early radar antenna. The Whitley had a single Vickers .303 gun in the nose turret but four .303 Brownings in the tail. With a maximum speed of just 222 mph it was an easy target for enemy fighters. (Simon Parry)

Poland's 304 Sq. RAF flew the Vickers-Armstrong Wellington, a unique bomber of geodetic design nicknamed "Wimpey" by her crews. Built in great numbers for Bomber Command, it was adapted to serve in the antisub role. (Jerry Scutts)

CHAPTER 2

WAR IN THE NORTH ATLANTIC

The shipping lanes from America to England were largely north of 40 degrees latitude and it was in these North Atlantic waters that most U-boat operations were conducted from 1939 through 1941. To the British these routes were known as the Western Approaches.

At first the east bound convoys were escorted by Canadian naval vessels to the limit of their range, then picked up by Royal Navy escorts in the Western Approaches. The gap between the relatively short range escorts was prime hunting grounds for German submarines. In May 1940, after the fall of Denmark and Norway, Danish-Iceland at the top of the North Atlantic route was seized by Britain both to deny it to the Germans and to provide a mid-Atlantic base for surface escorts and ASW aircraft.

As the war progressed, and Allied equipment improved and technology advanced, aircraft operating from Canada, Iceland, and the British Isles began to correlate more closely with surface forces. This left fewer voids along the convoy routes and forced the BdU to send their U-boats to the Americas and the South Atlantic. Still the majority of the submarine offensive was undertaken in the frigid waters of the North Atlantic, the main street of the supply line to England.

Both German and Allied code breakers began a round-the-clock game of decrypting each other's radio signals to divine the routes of convoys and the gathering locations of U-boats. As this technology advanced, the Allies had the better of it and would frequently thwart the assembly of German submarines.

Admiral Doenitz had evolved a strategy of forming a picket line of submarines across expected convoy routes. When one U-boat sighted the convoy, it would shadow while transmitting a homing beacon to the rest of the flotilla. Thus the wolf pack would gather for the kill.

After Hitler attacked the Soviets in June 1941, an even greater volume of ship traffic utilized the northern route, eventually sailing from England between Norway and the Arctic pack ice to the Northern Russian ports of Murmansk and Archangel.

As the struggle for control of the North Atlantic progressed, the presence of growing numbers of Allied ASW aircraft had an impact even if they could not sink U-boats. The sub hunters frequently compelled their quarry to dive long enough to permit convoys to escape.

Even in summer the weather over the North Atlantic was frequently appalling. It was not at all infrequent for patrolling aircraft to simply vanish over these vast reaches of ocean. Mechanical or navigational errors were as deadly as enemy action. The survival potential for seaman or airmen downed in these waters was dismal, ranging from minutes to a few hours.

A Catalina I, SN AM266, was tested as a torpedo carrier by 240 Sq RAF Coastal Command. This was the squadron that shadowed the German battleship, *Bismarck,* leading to her ultimate destruction. (David Legg)

A USN Consolidated Catalina PBY of VP-73 being refueled at Reykjavik, Iceland. Lieutenant R.B. Hopgood's aircraft from this unit scored its first victory on 20 Aug '42, sinking U-464 while covering a convoy south of Iceland. The Catalina was the best antisub aircraft available at the start of the Battle of the Atlantic, the Canadians and the British equipping many units with the durable flying boat. (NARS)

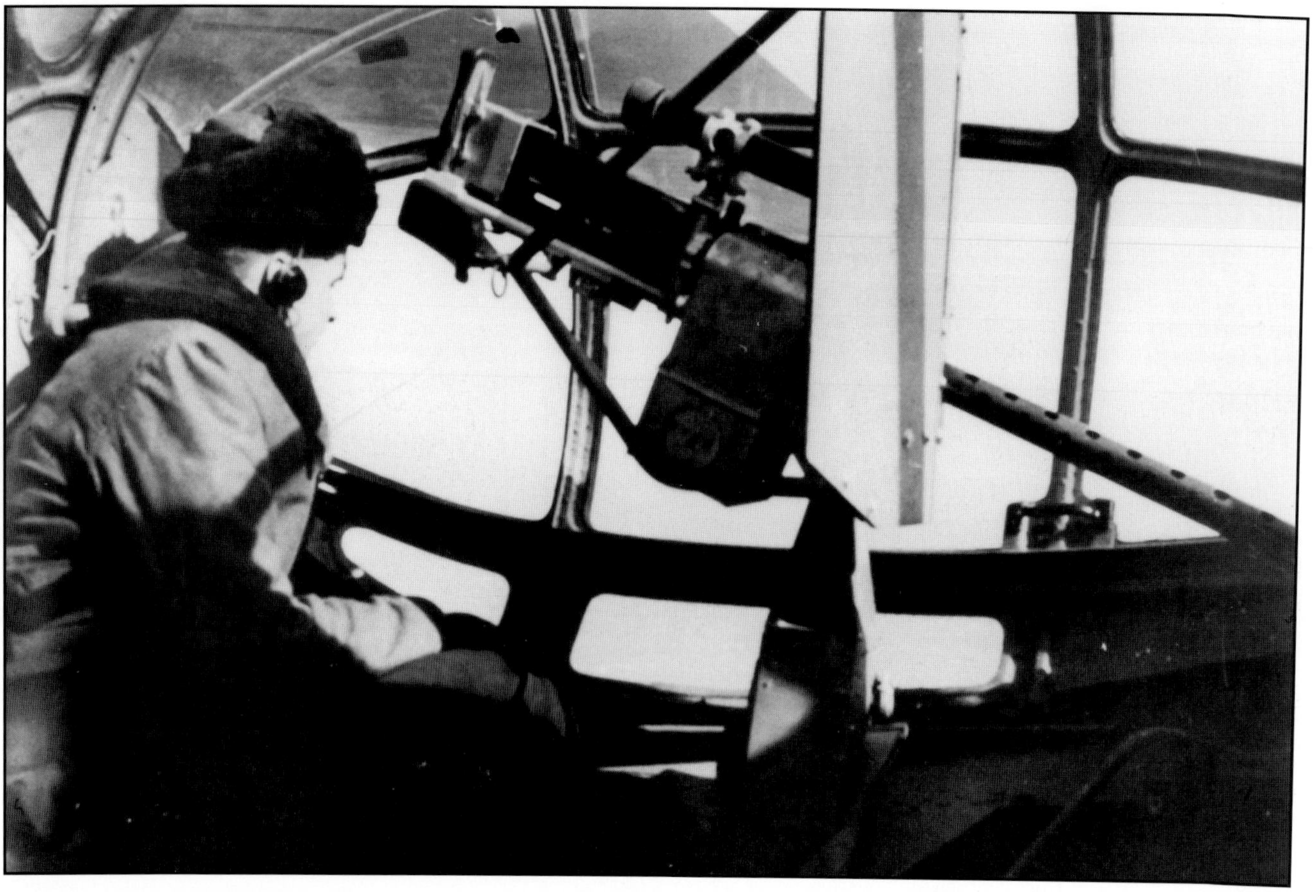

Above: One crew member of VP-73 mans a lookout post at the starboard gun blister of his PBY. The .50 caliber machine gun was used in strafing attacks and to defend against enemy aircraft. Below: A galley scene on a VP-73 Catalina. Hot food and coffee were essential for the ten hour missions in the sub-Arctic cold. (NARS)

Eventually new equipment began to augment and replace the early aircraft with RCAF. Above: A Lockheed Ventura of 145 Sq taking off from Torbay, Newfoundland, in Sep '43. The Ventura was a descendant of the Hudson. (National Archives of Canada) Below: Consolidated Liberators began to appear in 1943. The nose bulge on this 10 Sq Lib houses its radar. (Canadian Armed Forces)

The importance of remote Iceland in the Battle of the Atlantic was such that the AAF provided it with an air defense until 1944. On five occasions long-range German reconnaissance aircraft were intercepted and downed near Iceland. Above: Bell P-39D Airacobras of the 33rd FS are seen in their sand bag bunkers. Below: The Lockheed P-38F of Lt. Elza Shahan, 27th FS took some hits from the FW 200 that he and Lt. Joseph Shaffer, 33rd FS, bagged on 14 Aug '42. A shot from the Luftwaffe gunners entered the nose and landed in the P-38's ammunition magazine. (AF Coll.)

Above: The Focke-Wulf 200 long-range reconnaissance bomber was adapted from the Luft Hansa Condor airline transport. These planes roamed far over the Atlantic and the North Sea hunting convoys and cooperating with BdU. They attained such an impressive record of ship kills that Winston Churchill labeled them, "the scourge of the Atlantic." Frequently encountered by Allied aircraft, the huge ship would employ her six defensive machine guns offensively when warranted. Many operated from the Bordeaux area with KG 40. Below: This FW 200 was shot down while on recon over Iceland 24 Oct '42 by 2nd Lts. Michael Ingelido and Thurman Morrison flying Bell P-39D fighters of the AAF's 33rd FS. (NARS)

An RCAF 162 Sq late model Catalina built at Vickers in Montreal (SN 11090) warms up and lumbers into the air from Camp Maple Leaf, near Reykjavik, Iceland in 1944. Catalina crews joked, "It's not much of an airplane and a lousy boat." But the "Canso," as it was labeled by Allied airmen, was a stellar sub hunter because of its range, capacity, and durability. (Canadian Armed Forces)

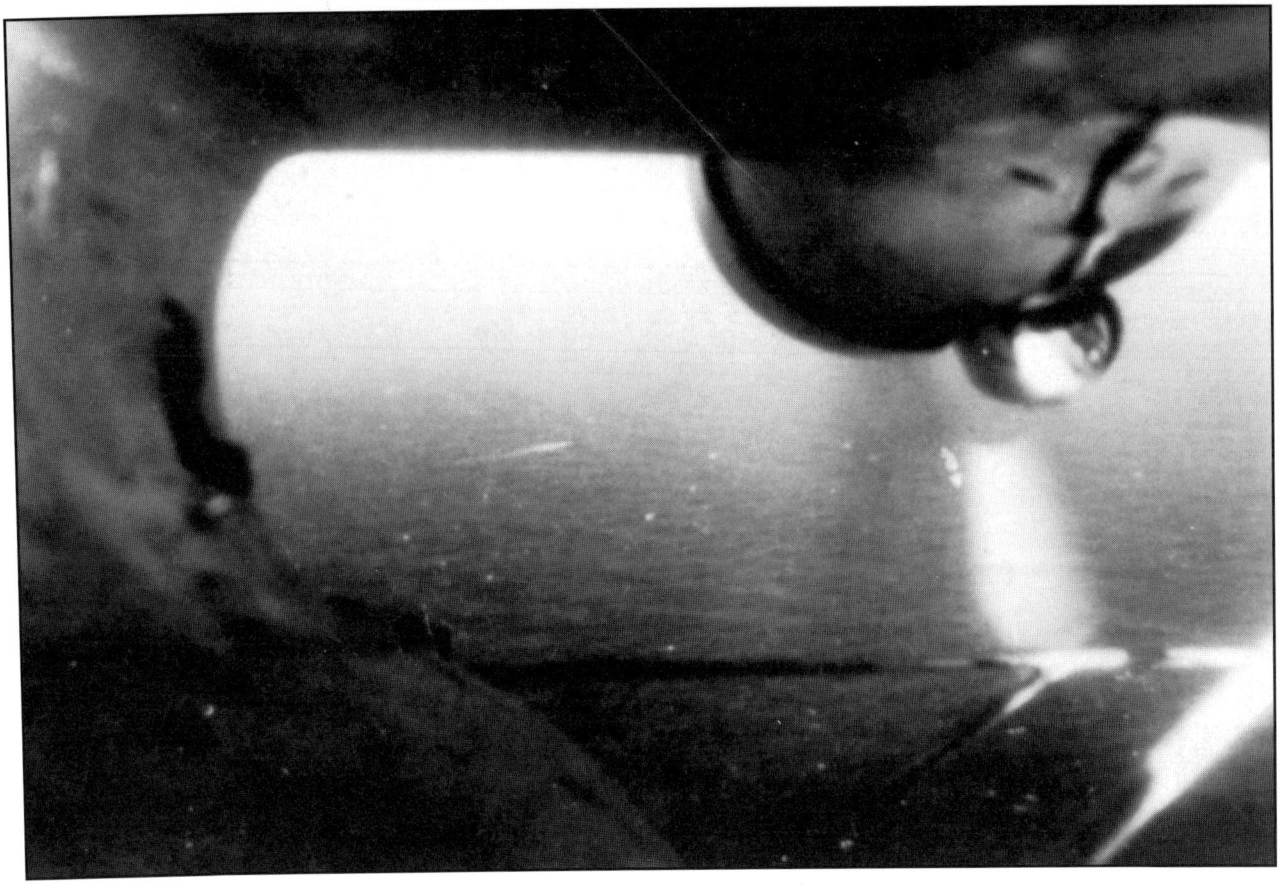

Above: A thin wake on the horizon reveals a surfaced U-boat in the North Atlantic as first seen from the right blister turret of Canso "W" of 5 Sq, RCAF. Below: Looking back, depth charges are dropped close by U-209. It sank with all hands on 4 May '43 S.E. of Greenland. (NARS)

Above: Some RCAF units were equipped with Short Sunderland flying boats. Here aircraft "Z" of 422 Sq is seen in Apr '43 operating from Northern Ireland. Below: Sunderland "U" of RCAF 422 Sq, shown on take-off from Castle Archdale, Northern Ireland, was commanded by W.F. Morton. (Canadian Armed Forces)

Above: Pictured on the ramp at Castle Archdale, Northern Ireland, in July 1942 are an RAF Catalina (left) and an RCAF Sunderland (right). The latter aircraft equipped RCAF Sqs 422 and 423. Below: RCAF aircraft mechanics at Castle Archdale had to don rubberized wading suits in order to cope with flying boats afloat in the icy waters of Lough Erne. (Canadian Armed Forces)

The "Leigh Light" was conceived by RAF Sq. Ldr. H. de V. Leigh in mid-1942. The advent of this ingenious gadget stripped away the U-boats' protection that had been afforded by darkness, a period when submarines routinely surfaced to recharge their batteries. The 24-inch, 70-million candle-power light was installed in various locations and is seen here mounted under the wing of an RAF Coastal Command Liberator. (Simon Parry)

No longer getting "hand-me-downs" from Bomber Command, by 1943 CC began receiving aircraft specifically developed for sub hunting. Shown here is a Vickers Wellington Mark XII, a state-of-the-art ASW aircraft with factory installed radar chin and retractable Leigh Light in ventral position on the underside of the fuselage just aft the wing. Range was 1,000 miles. It carried a pair of .303 guns in nose turret, single guns at either waist window, and four .303s in the tail. RCAF 407 Sq was equipped with the "Wimpey," as British aircrews had christened it. (Canadian Armed Forces)

In-flight shot of a 300 (Polish) Sq Vickers-Armstrong Wellington displays the chin modification that housed radar equipment. (Simon Parry)

The Vickers-Armstrong Warwick, a modified Wellington, had developmental problems, and limited numbers joined CC late in the war, serving in both ASW and rescue roles. This aircraft carries the markings of RAF 179 Sq which operated out of England and Gibraltar. (Simon Parry)

Another late arrival in Coastal Command's inventory, a Mark V, Handley Page Halifax of CC 518 Sq at an airfield in Northern Ireland in 1944. It was the aircraft of Sq. Ldr. Young. (Simon Parry)

CHAPTER 3

BATTLE OFF THE AMERICAS

By December 1941, Admiral Karl Doenitz had steadily advanced his submarines westward in the Atlantic. And although the U-boats had engaged and sunk U.S. merchantmen and even U.S. Navy escorts during the convoy battles, Hitler had not wanted to provoke America's entry into the conflict. Only a handful of attacks had been made against ships near the Western Hemisphere and those were off Canada.

The Japanese attack on Pearl Harbor, 7 December 1941, swept all that aside. Doenitz immediately ordered his submarines to patrol lines off the U.S. where they raised havoc, sinking 360 ships in the first six months of 1942.

The east coast of the United States was littered with the flotsam of lost ships and the fires of burning tankers could be seen from shore. U-boats invaded the Caribbean, sinking ships as far west as Mexico and in the approaches to the Panama Canal, and ranged to the South Atlantic off the coast of Brazil.

Although the U. S. industrial establishment had been gearing up for war for many months, it did not include weapons to counter a submarine offensive. U.S. Coast Guard cutters were refitted for ASW operations and the building of 180-foot patrol craft and 110-foot sub chasers went into high gear, many being built at inland shipyards. The Navy even dragooned private yachts into the ASW business.

Just as the British had done, the Army Air Force and Navy scraped together aircraft for off-shore patrol whose chief qualification was their availability. The only aircraft in the U.S. arsenal that resembled an anti-sub weapon was the Consolidated Catalina PBY flying boat. Douglas B-18 Bolos, known to be obsolete for any other combat, were quickly utilized as were handfuls of new North American B-25 medium bombers, the crews woefully ignorant of the task ahead. Single-engine AAF fighters and Douglas A-29s were flown within sight of coastal areas as were Navy Chance Vought Kingfishers and Coast Guard Curtiss Seagulls. The AAF went so far as to organize civilian aircraft into a Civil Air Patrol, sending Cessnas and Pipers out to look for submarines.

The U.S. Navy with its Catalinas and a few Lockheed Hudsons, began to string out small squadrons from Newfoundland to Bermuda. The first "kills" for U.S. aircraft were made on 6 Mar '42 by a Hudson (Ens. William Tepuni) and a Catalina (Chief Aviation Machinist's Mate Donald Mason) of VP-82 operating from Argentia, Newfoundland. Only a handful of additional U-boats were destroyed in American waters for 1942 by the combined efforts of ships and aircraft, such being the state of training, tactics, and technology. At the outset there was no airborne radar. U-boats were hunted with the "Mark I Eyeball." At this point, the battle of the Atlantic was decidedly being won by the Germans.

As the numbers and quality of aircraft grew, so did the string of air bases. Eventually they ran from the Canadian Maritime Provinces through the United States across the Caribbean and down through South America to Brazil. The B-18s and single-engine stopgaps were replaced with more of the marvelous Catalinas, and some suitable but less outstanding aircraft such as the Martin Mariner and the Lockheed Ventura.

By 1943 the happy hunting grounds off the Americas gradually became a graveyard for U-boats.

Early in 1942 the fires of sinking ships could be seen from the east coast of the U.S. Above: The SS *Dixie Arrow* was torpedoed off Cape Hatteras on 26 Mar '42 by U-71. Below: The Mexican tanker, *Portrero Del Llano* was torpedoed by U-564 near Miami on 14 May '42. This sinking hastened Mexico's declaration of war on the Axis. (NARS)

In Feb '42 the AAF, desperate for aircraft to patrol off the busy port of Miami, scraped together a squadron of obsolete North American O-47s, airplanes intended to operate with ground forces. They could scout but had limited range and little with which to fight a U-boat. (Sam Smith)

A Vought OS2U Kingfisher, normally an observation aircraft serving from battleships and cruisers, is seen here employed in the role of coastal convoy guardian off Virginia in Sep '42. On 15 Jul '42 the U-576 had torpedoed two ships off Cape Hatteras when it was discovered by a Kingfisher of VS-9. Summoning S.S. *Unicoi*, an armed merchantman, they collaborated to attack the sub which *Unicoi* finally rammed and sank. (NARS)

Loaded merchant ships from the Americas, bound for England and Russia, assembled into massive convoys at Halifax, Nova Scotia. There, escort vessels and air coverage were assigned. This is a view of Canada's Eastern Air Command plotting center in Halifax. German submarines hounded these eastbound convoys and even boldly entered the Gulf of St. Lawrence. (Canadian Armed Forces)

P/O W.B. Cooper and crew, 407 Sq RCAF, examine the flak damage to their Hudson's tail after an inconclusive engagement with a U-boat. Early BdU tactics directed U-boats to dive when aircraft were encountered. But these tactics gave way in 1942 to "stand and fight" orders. By 1943, many subs had been equipped with additional AA armament. (Canadian Armed Forces)

With the advent of the concerted German submarine campaign, the RCAF was suddenly thrust into a major role in defense of the vital Atlantic supply line. Initially they relied on whatever the American military could spare or the American aviation industry could supply. Above: An obsolete AAF Douglas B-18, dubbed "Digby" by the RCAF, of 10 Sq. Below: A Lockheed Hudson Mark I of 145 Sq in revised light paint scheme. (National Archives of Canada and Canadian Armed Forces)

With a two-ocean commitment, the U.S. Navy was desperate for patrol aircraft. They took early model Lockheed Hudsons from the British contract and designated them PBO-1. This one was supplied to VP-82 operating from Argentia, Newfoundland, where Ens. William Tepuni's plane scored the first U.S. ASW victory, sinking U-656 on 1 Mar '42. (NAM)

A Consolidated Catalina PBY-2 of the Naval Training Command being launched for sub patrol from the ramp at Jacksonville NAS, FL in May 1942. (NAM)

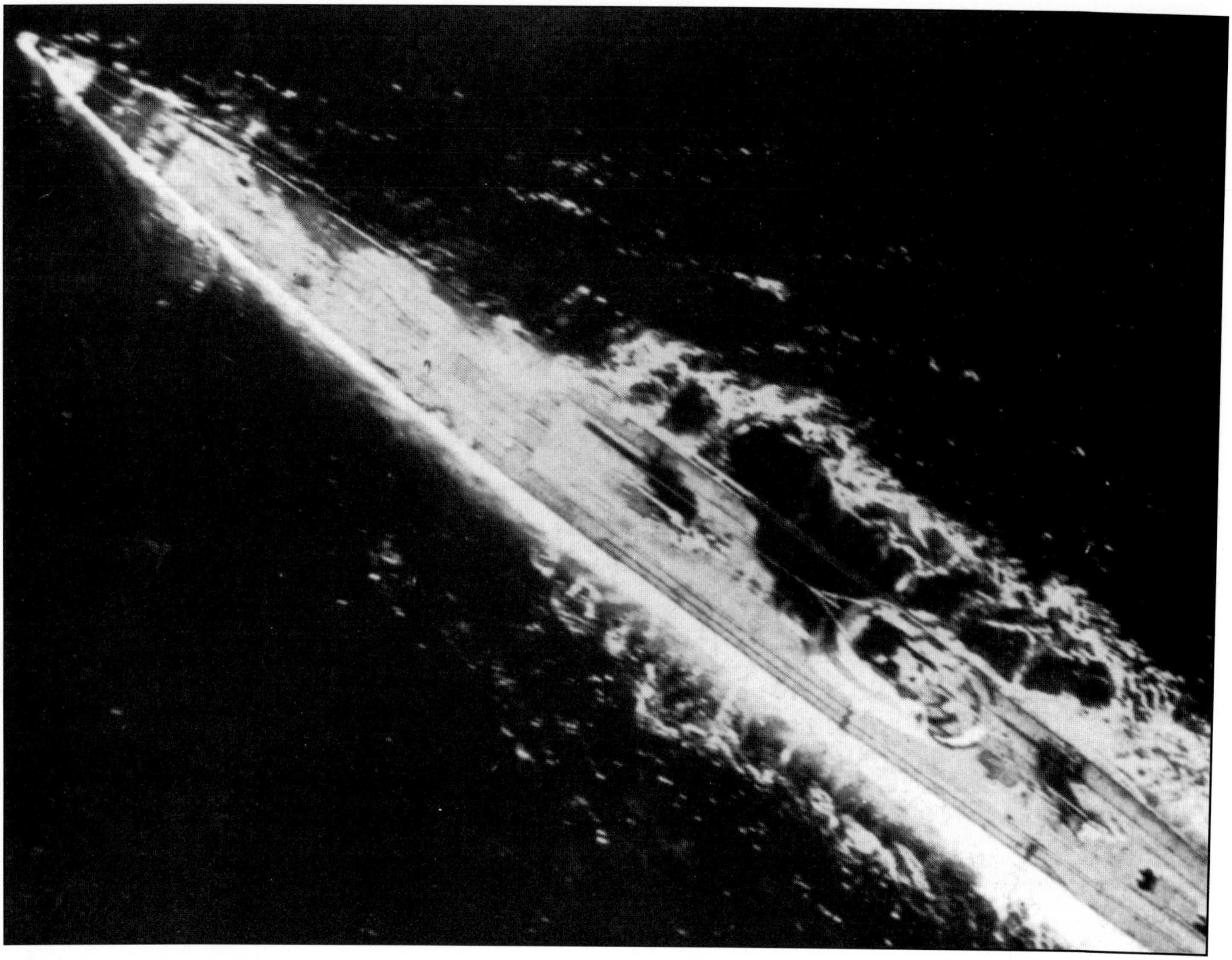

U-165 narrowly escaped destruction near Anticostia Is. at the Mouth of the St. Lawrence River on 9 Sep '42 by a Hudson of RCAF 113 Sq, Flt. Lt. R.S. Keetley commanding. The sub was later lost in the Bay of Biscay en route to its home port. (Canadian Armed Forces)

Based on a perceived British requirement, the Lockheed-Vega Ventura was developed as a light bomber (a derivative of the Hudson) but was deemed unsuitable for European operations. Because of its considerable range, it was "drafted" into the ASW business by both the Navy and other Allied air forces. This photo shows a PV-1 based in Argentia, Newfoundland with VP-82. Note that the top turret has been removed in this Dec '42 photo. (NAM)

The US Army Air Force employed the obsolete medium bomber, the Douglas B-18 Bolo in antisub patrols. They were useless for operations in regions where Japanese or German aircraft might appear but had the range and suitable bomb carrying capacity for ASW. Above: An unmodified version of the B-18 with its greenhouse nose at a Caribbean base. Below: This aircraft from the 10th AS Sq, British West Indies, has had the bombardier nose replaced with a radar bubble and a Magnetic Airborne Detector (MAD) inserted in the tail. (NARS)

The small U.S. Coast Guard air contingent had one notable success in the U-boat war. On 1 Aug '42 Chief Aviation Pilot Henry C. White and his radioman, George Boggs, were patrolling 100 miles off the Mississippi delta in a Grumman J4F-1 Widgeon when they sighted U-166 surfaced. An example of the Widgeon, with depth charge, is seen above. As White circled to attack from the stern the sub, apparently alerted, began to submerge. White dove the Widgeon at a fifty degree angle and Boggs saw the explosion of the single 325 lb. depth charge over the vanishing U-166. The direct hit sank U-166 with her entire crew. White was given a commission and both he and Boggs were decorated. Their actual aircraft (shown below) is now on display in its WW II markings at the Naval Air Museum, Pensacola. A postwar owner repowered the aircraft. (USCG)

A dissimilar pair of sub hunters. Above: The Coast Guard acquired a number of Curtiss SO3C-3 Seamews which were employed in ASW operations in both float and land configuration.These planes had a range of just over 600 miles. Developed as a replacement for the Curtiss Seagull biplane scout, the Seamew was underpowered and unloved by the Navy. (USCG) Below: The huge Martin PBM-3 Mariner had some early engine problems but helped fill a need in the Atlantic air war. Shown at take-off from Banana River, FL in Dec '42 is this VP-201 aircraft. Note search radar bulge above the forward cabin. The Mariner had a range of over 2,000 miles. (USN)

Lockheed PV-1 Venturas, a direct offspring of the Lockheed Hudson, were introduced into Navy ASW units in late 1942. They were bigger and superior in every respect to the old Hudson, and various modifications improved their range. Above: No. 125-B-6 of VB-125 at Argentia, Newfoundland, displays its under-wing auxiliary drop tanks. Below: This crew of VB-125 sank U-174 on 27 Apr '43 south of Newfoundland. (L. to R.) pilot Lt. (jg) Tom Kinasczczuk, co-pilot Ens. Robert Slagle, gunner Joe Holt, and radioman Robert Berg. A pair of fixed .50 cal machine guns can be seen above the nose. (NARS)

German attempts to interdict South Atlantic cargo bound for the Allies caused the sinking of several ships including some Brazilian merchantmen. This brought Brazil into the war on the side of the Allies and hastened the basing of U.S. Navy ASW units there. On 6 Jan '43 a Catalina of VP-83 (Lt. William Ford commanding) sank U-164 in the South Atlantic then threw a life raft to a handful of survivors. Two of them reached the north coast of Brazil and were made POWs. VP-83 transported the Germans (center foreground) to Natal. Note the pair of depth charges under each wing. (NAM)

Attacks by Martin Mariner PBM-3s of VP-74, commanded by Lt. H.S. Davis and Lt. (jg) Harold C. Carey, brought U-128 to the surface off Brazil on 17 May '43. DDs *Moffett* and *Jouett*, summoned to the encounter, finished the submarine with their guns. The PBMs then dropped life rafts to the forty-seven survivors. On 3 Jul '43 Carey and his crew were shot down and killed in an attack on U-199. (NARS)

An ASW version of the B-24 Liberator, adjudged the best all-round sub hunter, began to equip Navy units in 1943. Seen here is a Consolidated PB4Y-1 of VB-103 on patrol off Norfolk, VA just prior to deployment to England. (NAM)

Patrolling off the Atlantic coast of the U.S. in 1943 is a Navy antisub combo, a PBY-5A Catalina of VP-63 and a Navy Blimp. The ungainly airships had a top speed of just forty-seven miles per hour but had good range and the ability to virtually hover near suspect radar contacts. K-74, commanded by Lt. Nelson G. Grills, was doing just that about midnight on 18 Jul '43 and located U-134 off the Florida Keys. Grills attacked the sub without waiting for assistance and was shot down by the U-boat's AA fire. Nine of ten crew members were rescued the next day. Ultimately U-134 was sunk by a Wellington of 179 Sq RAF on 24 Aug '43 off the coast of Spain with the loss of all hands. (NAM)

Above: U-134, was what the Germans called a "milch cow," a replenishment submarine. It had downed airship K-74, and was observed east of Bermuda on 8 Jul '43 by a Martin PBM Mariner of VP-201. A/C commander, Lt. John T. Hitchcock brought his giant seaplane down to masthead height for an attack, but U-134 again escaped destruction. Right: Hitchcock's PBM took five hits from the sub's gunners and is shown on the ramp at Bermuda for repairs. (NARS)

A Canso of RCAF 161 Sq taxis onto the seaplane ramp at Gaspe, New Brunswick, Canada. (Canadian Armed Forces)

U-662 is viewed from the VP-94 PBY of Lt. (jg) S.E. Auslander off the coast of Brazil on 21 Jul '43. A squadron mate, Lt. (jg) R.H. Howland joined the hunt and sank the U-boat. Only three Germans submariners survived. (NARS)

Right: Cutting the waters of the South Atlantic at flank speed, U-161 is seen from a PBM-3 of VP-74 off Brazil. The Mariner crew, Lt. (j.g) Harry B. Patterson commanding, sank the sub with all hands on 27 Sep '43. (NARS)

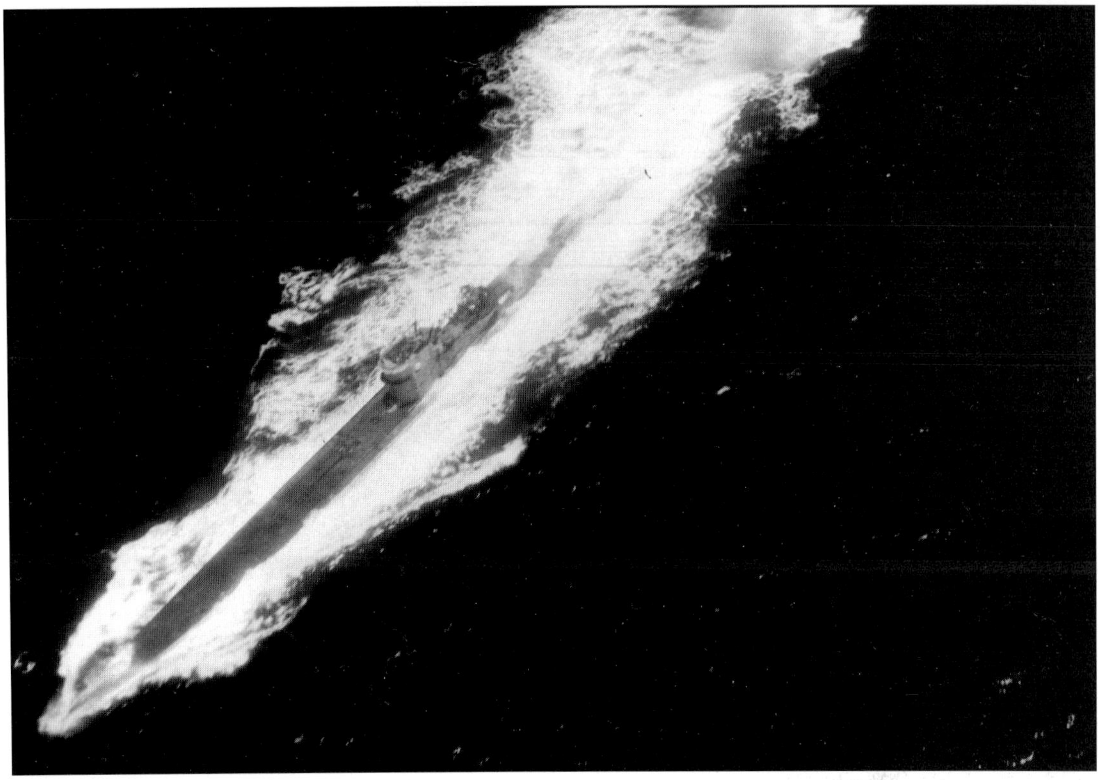

Below: U-849 was headed for the Indian Ocean to raid commerce when she was found off the African coast on 25 Nov '43 by an Ascension Island based Liberator of VB-107. Seen here the U-boat has just taken a lethal hit. Lt. (jg) Vance Dawkins delivered his bombs from such a low altitude that he received blast damage to a tail fin and rudder. None of the sub crew survived. (NARS)

U-848 is seen here on 5 Nov '43 under attack southeast of Ascension Island by VB-107. The PB4Y of Lt. William Ford has just passed over, and the explosion of its bomb can be seen in the geyser of water, smaller splashes marking the trail of .50 cal. bullets that appear to rake the superstructure. Charles Baldwin, William Hill, and Sam Taylor commanded the other VB-107 Libs that sunk U-848. It was Ford's third sub action. (NARS)

In late 1943, a VP-211 Martin PBM Mariner is seen in flight off Rio de Janeiro, Brazil. Sugar Loaf Mt. is in the background. (NAM)

Partners in antisub operations: USN Airship K-88 is shown moored to a movable mast at Santa Cruz, Brazil. In the foreground are Vultee Vengeance dive-bombers wearing the Brazilian insignia, used for coastal patrol. (NARS)

As the Brazilian armed forces took over more of their own coastal protection, the U.S. transferred ASW aircraft. In a ceremony at Ibura Field, Recife, the Brazilians unveil one of five Lockheed PV-1s on 30 Mar '44. The Brazilians also operated Catalinas. (NARS)

CHAPTER 4

THE BAY OF BISCAY

Submarine bases created at ports on the French coast facing the North Atlantic had been a boon to the German ocean offensive. However, as Allied forces grew in size and advanced in the development of ASW technology, the Bay of Biscay became a killing ground for U-boats.

From bases on the southwestern tip of England – St. Eval, Mount Batten, Dunkeswell, Chivenor and Hamworthy – RAF Coastal Command began patrolling the approximate 300-mile mouth of the Bay of Biscay in 1942. With the arrival of U.S. reinforcements in 1943, the air patrols became more numerous and German U-boat losses soared.

In November 1942, the AAF's 480th Antisubmarine Group brought their B-24 Liberators to England and began flying missions over the Bay of Biscay. The 479th AS Group followed, and by July 1943 the U.S. Navy's Fleet Air Wing 7 began similar operations with VP-63 (Catalinas) and VB-105, 103 and 110 flying PB4Y-1 Libs. Some of these units also operated from French Morocco patrolling both the Bay of Biscay and the Straits of Gibraltar.

Submarines departing France on war patrols and those returning were subjected to such a concentrated surveillance by combined RAF, AAF, and USN that passage across the Bay became the most dangerous part of a U-boat's voyage.

To counter the Allied air patrols the Luftwaffe established a force to intercept and engage the sub hunters. Kampfgeschwader 40 (later designated Zerstoerergeschwader 1) sent squadrons equipped with Junkers 88 long-range fighters to bases in Western France.

The presence of German aircraft added an element of danger to ASW operations unknown in other parts of the Atlantic. There has been no known compilation of losses in aircraft and aircrews suffered by the variety of Allied squadrons. However, the recent book, **Bloody Biscay**, Crecy, 1997, by Chris Goss, states that Gruppe KG 40 and ZG 1 scored "109 known kills matched to Allied losses." The same history shows that the German squadrons suffered 120 aircraft lost to all causes. Clearly, it was risky warfare for airmen of both sides.

The variety of perils connected with these eight to twelve hour missions were many: weather, mechanical failure, shortage of fuel, landing and take-off accidents, the ever dangerous defensive fire from U-boats, and interception by German aircraft. Yet one additional stroke of misfortune overtook a PB4Y from VB-110. The crew was killed on a night mission when they ran down a radar contact they thought was a U-boat. Their radar had acquired Skellig Michael Rock, a tiny outcropping off the Irish coast.

A month after the Normandy invasion of June 1944, the submarine ports in Western France were lost to Germany and this bloody chapter in the U-boat war was closed.

Engine trouble forced this RAF 502 Sq. Whitley to ditch in the Atlantic west of Lorient, France on 16 Nov '42. Note the crew in dinghy near the tail. A Polish destroyer rescued the six-man crew the following day. (PRO)

The importance of western French ports to the German Navy caused them to station special fighter units that could range over the Bay of Biscay to defend transiting U-boats and engage Allied sub hunters. The Junkers 88, designed as a light bomber, had the range and, with the removal of the glass nose and installation of three 20mm cannons and three machine guns, it became a deadly stalker of the sub hunters. Two German units, KG 40 and ZG 1, flying primarily the Ju. 88C-6, downed some 109 Allied aircraft between July '42 and July '44. Their claims were mostly for Liberators and Coastal Command Sunderlands, Wellingtons, and Whitleys. (NARS)

The earliest U.S. arrivals in England were AAF Liberators, stripped of the high altitude equipment being utilized by the 8th AF but provided with search radar. Above: TIDEWATER TILLIE, SN 42-40329, was a B-24D (Roy J. Hopper). She was replaced by a new version (below) with a chin modification housing radar. TILLIE, shown here at Dunkeswell, England, was assigned to 2nd Sq, 480th AS Grp. She had a record of several engagements under her pilot, Lt. William L. Sanford, and sank two subs: U-519 on 10 Feb '43 and U-524 on 22 Mar '43. (Al Lloyd)

A U-boat is captured in the Atlantic west of France by the camera of a St. Eval's Liberator, but not sunk. Note the German gunners manning their cannon. (NARS)

WHITE SAVAGE was a 479th AS Grp B-24D, SN 42-40921, specifically modified with a Convair A-6 nose turret and droop chin for antisub operations. (Al Lloyd)

Above: B-24D, SN 41-24002, SKEEZIX, 1st Sq, 480th AS Grp, had a second name, DAVEY JONES' HELPER. This photo shows the early hand operated .50 cal. nose machine guns. The Libs of the under strength 480th were stripped of some standard factory installed gear for high altitude operations and fitted with extra bomb bay fuel tanks. (Al Lloyd)

Below: CHARLOTTE THE HARLOT was another B-24D, SN 42-40339, of the 480th AS Grp. (Al Lloyd)

On 5 Jul '43 west of Lisbon, Lt. Carl G. Damann's 2nd Sq, 480th AS Grp B-24 detected this U-boat on radar and ran down the bearing. They attacked but the bomb bay doors refused to open and both hunter and hunted lived to fight another day. Damann and four other crew members were killed on a night landing the following month. (Roy J. Hopper)

1Lt Walter McDonnell and his B-24 crew from 1st Sq, 480th AS Grp, sank U-951 on 7 Jul '43. Just four days later they caught another U-boat (above) on the surface west of French Morocco but their depth charges fell short. Note the elevated AA platform behind the superstructure on both photos. (NARS)

A B-24 of the 480th AS Grp encounters a German Focke Wulf 200K Condor over the Bay of Biscay and trades shots. This was just one of dozens of engagements between Luftwaffe and ASW aircraft over the Bay of Biscay and the North Atlantic. (Roy J. Hopper)

This Liberator of the 480th AS Grp ran low on fuel and landed on the hardpan desert floor of French Morocco in March 1943. A "D" model, its SN was 41-23872 and it was named UBANGI WEBANGI. The shark's mouth was unique among 480th B-24s. (Roy J. Hopper)

B-24 "Q" of the 480th AS Grp, Capt. Lawrence E. Jarnagin pilot, sweeps in low to attack U-642 on 10 Jul '43. The Lib's crew made two runs but the U-boat skillfully eluded the Mark 47 depth charges. On 27 Aug '43 Jarnagin and his entire crew were lost off Cape Finisterre, Spain, in an aerial engagement with a Ju. 88 of 15/KG 40. (NARS)

2nd Lt. Ernest Salm and his B-24 crew from 1st Sq, 480th AS Grp have just passed over U-506 west of Portugal and photographed the explosion of their depth charges that would sink the sub on 12 Jul '43. Salm dropped a life raft to survivors, six of whom would be rescued by a British DD. (NARS)

Left: 30 Jul '43 off Spain, 1st Lt. A.L. Leal's B-24 of the 19th Sq, 479th AS Grp, attacked U-461 but failed to score. The depth charges, which would fall long of the target, can be seen near the center. The first bomb has just splashed. Three weeks later Leal's aircraft was damaged by a Ju. 88, but he brought it back to England. (NARS)

Lt. Earl A. Powers was pilot on MAID MARIAN, a B-24D of 2nd Sq, 480th AS Grp. The 480th began operations at St. Eval, England but transferred to Craw Field, Port Lyautey, Fr. Morocco as the invasion of Sicily attracted U-boat traffic toward the mouth of the Mediterranean. This shot provides a good view of the three flexible .50 cal. nose machine guns. Along with the top turret, nose guns would strafe a U-boat during the bomb run. (Roy J. Hooper)

Above: Roy Hooper's crew from the 2nd Sq, 480th AS Grp stands for their portrait at Craw Field, Fr. Morocco. These crews flew patrols as far as 1,200 miles for up to fifteen hours. The 480th's losses to all causes was 55%. Right: The aircraft in both photos are B-24Ds flown by another crew. The one with the nose art is SN 41-24196, a replacement. The sub silhouettes indicate eight sightings and three probables. Added later were VARGA GIRL and the swastikas indicating one German aircraft victory and three probables. Wounded in combat, VARGA GIRL also got a Purple Heart decoration. (Roy J. Hooper)

Left: A Consolidated Liberator PB4Y-1 of VB-103 over the coast of Southwestern England late in 1943. This particular aircraft, B-3, was lost on 12 Nov '43 in what must have been a battle to the death between U-508 and Lt. (jg) Ralph Brownell's Lib. Brownell signaled that he had made contact with a U-boat but never returned to Dunkeswell. BdU determined that U-508 was lost with all hands at this time in the Bay of Biscay. (NAM)

Below: A Mark III Sunderland of RAF No. 4 OTU shows off her armament. The nose turret generally carried two .303 machine guns but some were fitted with four. The top dorsal turret had a pair of guns and the tail turret mounted four .303s. Single Browning .50 guns were available for the side hatches just under the wing. This array of up to twelve machine guns earned the Sunderland the nickname, "Flying Porcupine" from the Luftwaffe. (Jerry Scutts)

Right: This attack by the Liberator of Lt. George Enloe, VB-103, west of Ireland on 28 Jan '44 resulted in the sinking of U-271 and all her crew. (NARS)

Below: An early model PB4Y-1 Liberator (B-24D) of FAW 7 stationed in Southwest England. The words on the nose are "Tally Ho! The Fox! She Runs!" (NARS)

By Dec '43 FAW 7 had improved versions of the PB4Y-1 in England. A VB-103 Lib with the Emerson nose turret is shown here. (NAM)

Lt. Oren Clark and his crew of PB4Y-1 "J" of VB-114 musters just prior to a night mission from England. They were on temporary assignment from Port Lyautey, Morocco, in 1944. (NAM)

This 304 (Polish) Sq Wellington received major damage from an encounter with a U-boat over the Bay of Biscay. The sub also survived this 5 May '44 action. (Simon Parry)

Navy's Liberator "O" of VB-110 patrolling off the southwest approaches to England in Sep 1944. (NAM)

Left: PB4Y-1 FEARLESS FOSDICK of VB-103 at Dunkeswell, England. Don Schierenbeck, co-pilot, with forty-three missions, peers from the starboard window. He previously flew ten missions in Catalinas with VP-73 from Agadir, French Morocco. (Don Schierenbeck)

Below: The crew of VB-103's FEARLESS FOSDICK near the end of their combat tour. (L. to R.) Top: Sherwin, Rayner, Windall, and Don Schierenbeck; middle: Berger, Dereese, Charillo, Garneau; bottom: Clark, Collier, the mascot named "Bombsite", Kurfiss, and Gilbert. (Don Schierenbeck)

CHAPTER 5

THE ESCORT CARRIERS

The early carnage created by U-boats had prompted the U.S. Navy to rethink its antisubmarine warfare doctrine. In late 1941, the Royal Navy's small aircraft carrier *Audacity* began working with an escort force on the U.K.-Mediterranean run, operating in close proximity to the merchant convoys. Her aircraft sank five U-boats before one got a torpedo into *Audacity* ending her career. But the lesson was clear. The glory of the escort carrier was her capacity to employ aircraft in mid-ocean where shore-based patrol planes could not reach and the ability to hound a submarine once it had been detected.

The Royal Navy built additional small "escort" carriers and the United States embarked on a massive construction program for the so-called "jeep" carriers or "baby flat-tops." The first of this new type vessel, the *Bogue* (Class) was built from converted freighter hulls. Eventually the U.S. Navy received another batch from Kaiser Shipyard for service in both the Atlantic and Pacific.

While the CVEs varied somewhat in design and machinery, they averaged about 500 feet in length with a flight deck of about 400 feet. Each carrier generally carried a complement of some twenty to twenty-five aircraft. In the U.S. Navy they were organized as a composite squadron (VC-) and were equipped with the Grumman Avenger and the Grumman Wildcat.

Operating first with convoys of merchant vessels, and then ranging alone with their escorts, the small carrier task forces persecuted the U-boats, hunting them down with a combination of deadly air and surface tactics. Once the airborne VC units had detected a submarine they would summon some of the half-dozen DEs that accompanied the task force. Coordinating air and sonar surveillance they combined in a relentless pursuit. The air tactics involved a team of a Grumman F4F Wildcat and a Grumman TBF Avenger. When the quarry was sighted, the Wildcat pilot would strafe with his six .50 caliber machine guns to neutralize the sub's AA while the Avenger crew would make a bombing run.

Flying from any aircraft carrier was perilous, but the short decks and small hulls of the CVEs added an extra degree of anxiety to the profession of naval aviation. In the book, **Wildcats to Tomcats***, (Specialty Press, 1996) Ens. (later Captain) Zeke Cormier describes a not untypical landing approach to the bobbing CVE *Card*, after four prior passes:

> "On the next go-rounds I really concentrated, but witnessed a series of flight deck undulations which presented first, a plan view of the entire deck, followed by an alarming view of the "spud locker" at the stern — including the exposed ship's propeller. Poor little CVE-11 was pitching to the extent she took green water over the bow.
>
> Passes five and six were okay, but I got no 'cut' signal from the LSO because of the fouled deck.
>
> By now it definitely was getting dark, which seemed to produce a deleterious effect on my throttle-control technique. Thus, passes seven and eight were mere flybys. Now it was not only for-real night time, but the folks on board were growing tired of Ensign Cormier's one-man air

show. Rumor had it that the gunnery department was alerted to uncover the five-inchers in case they had to bring me down the hard way. Embarrassed beyond all hope, I sulked in my Avenger cockpit, imagining some German submariners out there, laughing their asses off.

The LSO, Ed Deacon, informed me that *we are* going to land on the next pass; wind, water or night be damned! He sweet-talked me into the best trap I made during the entire cruise and has been my hero ever since."

The enormous success of the CVEs was only slightly countered by the U-boats with the sinking of *Block Island* CVE-21 on 29 May '44 near the Cape Verde Islands. U-549, which had achieved this victory, was hunted down and sunk by the carrier's escorts.

Over a two year period the Jeep carriers of the USN were involved in the sinking or capture of fifty-six German U-boats.

*** Written with co-authors, Wally Schirra, Phil Wood, and Barrett Tillman.**

Hundreds of the stubby winged Grumman Wildcats were purchased by the Royal Navy which dubbed the aircraft "Martlet." This Mark I, with four .50 cal. machine guns, did not have the USN folding wing feature. (Simon Parry)

One of the weapons that revolutionized the war against the U-boats was the escort carrier. In 1943 the "jeep" carriers began operating in the Atlantic employing their aircraft as "hunter-killer" teams. No longer did Allied warships wait for the sounds of U-boats on their sonar or the sighting of torpedo wakes. The CVE air groups hunted for German submarines and attacked them at a distance from convoys. Here the U.S. Navy *Card*, CVE-11, is seen with a deck load of Grumman Avengers and Wildcats. With flight decks just over 400 feet in length, the CVEs were mini-aircraft carriers. (NARS)

Santee's air group (CVE-29) prepares for flight operations in Nov 1943. No. 5 and the last row of aircraft are VF-29 Wildcats and all other aircraft are TBF-1C Avengers. *Santee's* airmen sank three U-boats. (NAM)

Early in 1944 the *Guadalcanal* CVE-60 takes water over the bow in mid-Atlantic. Flight operations were treacherous but sometimes unavoidable under such conditions. A pair of Avengers, wings folded, are lashed to the deck. (NARS)

A TBF-1C Avenger of VC-9 prepares for launching from escort carrier *Mission Bay* CVE-59 in Jan 1944. The principal ASW weapon of the escort carriers, it had the capacity for 2,000 lbs. of bombs and carried four 3.5 inch rockets under each wing in later versions. (NAM)

The other half of the CVE's aerial team was the dependable Grumman Wildcat. The little fighter was uniquely suited to the short decks. Originally produced at the Grumman "Iron Works" the need was so great that an improved, licensed version was made by General Motors, the FM-1. Shown here is an aircraft of VC-55 from USS *Card*. The Wildcat had six .50 cal machine guns with which its pilot would strafe a U-boat while the Avenger conducted its bombing run. (NARS)

Above: A load of anti-sub ordnance in the bomb bay of a TBF Avenger on carrier *Guadalcanal* in Oct '44. On the left are a pair of depth charges. Right is an acoustic torpedo known by the nickname, "Fido." (NARS)

Left: A sonobuoy, dropped by parachute from Avengers helped relay sound signals on submerged U-boats to patrolling aircraft and surface ships. (NARS)

Aircraft from carrier *Bogue* spray this unidentified U-boat with machine gun fire in May '43 but could not sink her. (NARS)

A Grumman F4F, tail hook down, gets a wave-off from the LSO on *Card* in early 1943. (NARS)

In mid-Atlantic, on 22 May '43, VC-9 Avengers from *Bogue*, CVE-9, caught U-569 on radar and attacked. Lt. (jg) William F. Chamberlin dropped four depth charges (one visible, right center) and the sub dove. When, because of damage, it resurfaced a short time later, Lt. Howard Roberts was waiting and dropped another spread of depth charges. The sub dove again but had been fatally wounded and was obliged to surface. Twenty-four members of U-569's forty-three man-crew abandoned ship as the Canadian destroyer *St. Laurent* approached. This was the first "kill" for a U.S. escort carrier air group. Chamberlin was awarded a Silver Star and Roberts a DFC. (NARS)

U-118, a 1,600-ton tanker ("milch cow") with 20mm AA mounts had fought off air attacks early in June 1943, but her subsequent radio traffic was intercepted and *Bogue's* aircraft hunted her down west of the Canary Islands. Above: Caught on the surface by Avengers and Wildcats of VC-9, its deck crew can be seen clearly as a depth charge splashes close alongside. Below: The frantic attempts of the sub to avoid its attackers are apparent from its tortured wake. No less than seven *Bogue* aircraft participated in the hunt on 12 Jun, and it is believed that Lt. (jg) Harry E. Fryatt dropped the fatal bomb. Avengers dropped life rafts to some sixteen survivors, who were later rescued by a DD, but forty-three of U-118's crew were lost. (NARS)

A TBM-1C of VC-58 terminated this landing in the starboard catwalk of *Guadalcanal* CVE-60. (NARS)

U-66 was the first German sub engaged by aircraft of *Card* CVE-11. Ens. Richard L. Cormier got this view on 3 Aug '43, east of Bermuda, when he dove to the attack accompanied by Ens. Arnie Paulson in an escorting Wildcat. Cormier's depth charges failed to release on the first run and, despite heavy AA from the sub, he made a second run. Two bombs exploded doing considerable damage, but U-66 survived the attack and served until her sinking on 6 May '44. (USN)

Above: Lt. Cdr. Charles W. Brewer, led VC-13 on *Core* CVE-13 during many sub actions in 1943 and was personally involved in the sinking of U-487 and 378. He served another tour of duty in the Pacific in 1944 where he became an ace. Cdr. Brewer was killed in action over Saipan as CO of VF-15. (NARS)

Above: Lt. (jg) William F. Chamberlin of Aberdeen, WA was involved in the sinking of U-860 while piloting a TBF with VC-9. In between these actions he survived a crash-landing. (See page 95.) He was killed in action during the U-860 engagement by delivering his bomb load from such a low altitude that his Avenger was engulfed in the explosion. His decorations included the silver Star, DFC with Oak Leaf Cluster, and England's Distinguished Service Cross. (USN)

Right: *Card's* VC-1 team of Lt. Asbury H. Sallenger in a TBF and Ens. John F. Sprague, (pictured here) flying a Grumman F4F Wildcat located U-760, U-664 and U-262 refueling in the Bay of Biscay on 8 Aug '43. Sprague raked the three vessels with machine gun fire even as the Germans chose to fight it out, while Sallenger conducted a depth charge attack. The combined flak downed both aircraft. Sallenger and his gunner survived to be rescued the following day but his radioman, John Downes, and Sprague were both killed. Their action damaged U-262 which was forced to abort. U-664 was hunted down and sunk the following day. (NARS)

The U-boat's deadly array of defensive anti-aircraft guns is exhibited in this photo of U-550, forced to the surface on 16 Apr '44 off Nantucket by DEs. Sinking by the stern, she surrendered. Some of her crew prepare to abandon ship. A pair of 20mm dual cannons can be seen center and a single 37mm cannon aft on the "bandstand" that was retrofitted to many German subs in 1943. The 37mm, though powerful, tended to jam after one round. (NARS)

On 8 Jun '43 a pair of *Bogue* Avengers from VC-9 ran down a Huff-Duff bearing that led them to U-758 in mid-Atlantic. The sub had recently been refitted with quad 20mm cannons on a "bandstand" behind the conning tower and she chose to fight it out with Lts. (jg) Letson Balliett and W. S. Fowler. Their low level attacks severely damaged but did not destroy the sub. Fowler and his crew took five hits from the 20s, one as seen here on the starboard bomb bay door, but managed to return to their carrier. The aviators served a second tour of combat on *Card* and survived other encounters. (NARS)

Above: U-664 was tracked down on 9 Aug '43 by *Card* aircraft. A combination strafing-bombing attack by Lt. Norman Hodson in an F4F and Lt.s (jg) Gerald Hogan and R.H. Forney in Avengers forced the sub to the surface. Dead in the water, sinking by the stern, her crew began to abandon ship. Below: In this close-up the crew can be seen on the superstructure wearing life jackets. Note the ship's insignia, a smiling sawfish. (NARS)

An extraordinary scene, milch cow, U-117, (left) and U-66 (right) taken from an Avenger on 11 Aug '43 in mid-Atlantic. Ens. Richard L. Cormier's TBM attack on 3 Aug had damaged U-66, killed a gunner, wounded eight other crew, and seriously injured the captain. The Allies intercepted radio transmissions for a resupply by milch cow, U-117. *Card's* aircraft of VC-1 made the rendezvous, and the attack is shown here as the two subs parted to get up speed. As Wildcats strafed the decks, Lt. (jg) Asbury Sallenger dropped two depth charges close by U-117 and she began erratic maneuvering although defending with her AA. U-66 dove and U-117 began to settle by the stern. Lt. Charles Stapler and Lt. (jg) Junior Forney followed with acoustic torpedoes (dubbed "Fido"). One or more of the torpedoes found U-117 which sank along with sixty-two crewmen, but U-66 survived to reach France. The end of the road for U-66 came on 6 May '44 in a combination of carrier aircraft and a destroyer-escort action. Sallenger had been downed in a battle with U-boats only three days earlier. (USN)

One of the perils of the short-decked carriers was that aircraft occasionally missed the arresting wires with dire consequences. Above: A *Core* Avenger lands long but the barrier prevented major damage or injury. (NARS) Below: This *Bogue* TBM-1C crashed through the barrier and into a deck load of aircraft. The pilot Lt. (jg) R.T. McGusty and his VC-42 crew were uninjured. (NAM)

During a dawn patrol north of the Azores on 4 Oct '43 Lt. (jg) Robert Stearns from *Card* sighted a "school" of four surfaced subs, one a tanker. He radioed his find then strafed all four and concentrated his bombs and a Fido sinking the "milch cow," U-460, seen here running for its life. Several additional flights of Wildcats and Avengers from VC-9 soon had the remaining U-boats under strafing and bombing attack and Lt. (jg) Stewart Holt's Avenger sank U-422. (NARS)

In mid-Atlantic on 13 Oct '43, U-402 was shadowed by a team from the carrier *Card.* Lt. Cdr. Howard Avery and Ens. Barton C. Sheela, of VC-9, collaborated on the sinking. (NARS)

Typical ASW aircrews (from VC-13 on carrier *Core*) Above: A Grumman TBF-1 and crewmen (L. to R.) Bondy, Ens. Humphrey, and Slonger.

Left: The pilot of a Wildcat, Ens. Eares. Note the insignia on the side of the plane, a marking unique among CVE squadrons. (NARS)

U-378 was found and attacked by a Grumman Wildcat and Avenger team from *Core*, CVE-13, on 20 Oct '43 north of the Azores. Cdr. Charles W. Brewer, skipper of VC-13, flying the Wildcat, strafed the German gunners so effectively that Lt. R.W. Hayman's TBM got close enough for this picture as he straddled the sub with his depth charges. U-378 sank with all hands. (USN)

U-405 was one of a wolf pack of U-boats detected by *Card's* VC-1 aircraft on 30 Oct '43. When the Avengers had secured for the night, DD *Borie* took up the hunt on sonar and drove U-405 to the surface with depth charges. There ensued a running surface battle between the lone sub and the destroyer that involved an epic duel by deck guns, ramming, close quarter battle with small arms, and the eventual sinking of both ships. *Borie* is seen on fire and sinking by the stern, her casualties: twenty-seven officers and men. None of the U-boat's forty-nine crew survived. (NARS)

On 12 Dec '43 Lt. (jg) Elisha Gaylord, flying from *Bogue,* sighted and attacked U-172 west of the Canary Islands. Though damaged, the sub dove as Gaylord summoned other aircraft and ships. Despite severe depth charge attacks, U-172 managed to elude her tormentors until the following dawn. A patrolling TBF from VC-9 spotted the damaged sub's oil slick and the hunters renewed the attack. By the fourteenth depth charge attack U-172 was forced to surface and is seen here (lower center) smoking badly but still full of fight. The DD *Osmond Ingram* approaching U-172, raked her with gunfire and that finished the contest. Thirteen of the German crew members were killed but forty-six survived to be rescued. (NARS)

Lt (jg) Wallace LaFleur of *Bogue's* VC-19 was aloft in his Avenger West of Madeira on 20 Dec '43 when he sighted the surfaced U-850. His depth charges missed the target, but he summoned other TBFs and Wildcats from his squadron and, in subsequent attacks, they combined to wound and then sink U-850. The photo, taken from the Avenger of Lt. (jg) H.G. Bradshaw, shows one bomb raising a geyser well past the sub, but another has splashed close to the hull forward of the deck gun. There were no survivors. (NARS)

Operations from an aircraft carrier are high risk endeavors. Above: A catapult failure caused this Grumman TBF Avenger to be merely lobbed over the bow of *Card* on 28 May '43. The pilot, Lt. (jg) H.K Hass, can be seen standing in the forward cockpit and the lone radioman, W.D. Painter, is emerging from the hatch in the belly just visible under the right horizontal stabilizer. The forward momentum of the ship caused it to overrun the wreck and only Painter survived. Below: On 25 Mar '44 Lt. (jg) William Chamberlin approached his landing on the *Solomons* with too little power, resulting in a shattering crash against the stern ramp of the flight deck. (Note the LSO at the right.) Incredibly, all of the crew of this VC-9 Avenger were saved by the DD *Eisner*. Chamberlin, on his second tour of ASW duty, returned with a minor head injury. On 15 Jun '44, in the South Atlantic, Chamberlin was part of a VC-9 team pursuing U-860. He delivered his bombs at such a low level that his Avenger was engulfed in the explosion and all were lost. (NARS)

In yet another opportunity for disaster, a TBM-1C of VC-8 lost power on take-off from *Guadalcanal* and splashed. Pilot, Ens. B.J. Hudson, Radioman R.J. Horacek, and Gunner P. Schwahn were all rescued by an escorting DE. (NARS)

Atlantic fleet Avengers added rockets to their weaponry late in 1943. U-758 is seen south of Newfoundland on 11 Jan '44 just after Lt. (jg) Willis D. Seeley, from *Block Island* , had delivered the first rocket attack. Despite continued attacks by VC-58's Seeley and Lt. (jg) Leonard McFord, the sub managed to survive but was so damaged that it aborted patrol and returned to St. Nazaire. (NARS)

Above: A fatal attack on U-544 northwest of the Azores on 16 Jan '44 as seen from a TBM of VC-13. One depth charge fell long, but another exploded near the sub's center, sending it to the bottom with all hands. (NARS)

Right: Capt. Dan Gallery, skipper of the *Guadalcanal*, hears the tale of the sinking of U-544. (L. to R.) Lt. Cdr. A.H. Perry, CO of VC-13, Gallery, Ens. William H. McLane, Ens. Bert Hudson, and enlisted crewmen, R.J. Horacek, K.G. Dodd, P. Schwahn, and A.C. Torres. (NARS)

Above: U-515 was first sighted by an Avenger from *Guadalcanal* CVE-60 on 8 Apr '44 northwest of Madeira Island. A cat and mouse game ensued through that night and the following day with four Navy DEs joining the hunt. They delivered a pattern of depth charges that forced U-515 to the surface. Here she is seen broaching. VC-58 aircraft were there to pounce on the damaged sub and two DEs raked it with gunfire. Forty-three of the U-boat's crew were able to abandon ship and sixteen were lost as she sank. (NARS)

Below: Perhaps the best known photo of the U-boat war is this shot of captured U-505 alongside carrier *Guadalcanal*. The sub was returning from a patrol of the South Atlantic when her radio communications disclosed her relative position. Aircraft from *Guadalcanal's* VC-8 dropped sonobouys nearby that helped track the target west of Dakar on 4 Jun '44. Once captured on sonar by the DE *Chatelain*, other DEs gathered to deliver a depth charge attack that forced the sub to the surface. Too damaged to dive or run, U-505 surrendered and was quickly boarded by a prize crew. It yielded a treasure trove of intelligence material. Towed across the Atlantic and repaired, it rests today on dry land as a permanent exhibit at the Chicago Museum of Science and Industry. (NARS)

Aircraft from VC-12 off *Card* CVE-11 located U-233 in the sea lanes east of Halifax and she was forced to the surface then sunk by DEs *Baker* and *Thomas* on 5 Jul '44. Many of her crew were rescued including the mortally wounded skipper, Hans Steen. He died the following day aboard *Card* and was buried with full military honors. Here his body (in white canvas bag) is being committed. (NARS)

A primary factor in the success of the hunter-killer groups was the close coordination between aircraft and surface warships. TBM-3 Avengers of VC-12, USS *Core* CVE-13, overfly the DE *Loy* during a dusk patrol in the North Atlantic, Apr '44. (NARS)

Above: U-1229, one of the new snorkel equipped subs (the tall thick mast), was bound for Maine to land a spy when detected by the airborne radar of Lt. A.X. Broxas of VC-42 southeast of Newfoundland. He raced down the bearing to deliver an attack that caused severe damage to the sub. It made a shallow dive to escape, but Broxas kept his Avenger over the scene until reinforcements from *Bogue* arrived, two more TBMs and a pair of FM-2s. (NARS)

Below: The damage to the U-boat became such that it had to surface and VC-42 aircraft renewed the attack with bombs, rockets, and machine guns. Seen here from the Avenger of Lt. (jg) Bernie Sissler on 20 Aug '44, U-1229 is dead in the water and sinking as the crew abandons ship. Seventeen of her crew were lost but forty and the spy passenger were rescued. (NAM)

The Royal Navy operated many escort carriers primarily to protect North Atlantic convoys. Seen here in 1944 is *Campania* with her compliment of Fairey Swordfish and Grumman Martlets. (NAM)

The Fairey Albacore, "Applecore" to its crews, was intended (but never succeeded) as a replacement for the Swordfish. Built in smaller numbers, it was designed primarily for torpedo-bombing but also served in the antisub role. (NARS)

The Fairey Barracuda was also intended as a Fleet Air Arm replacement for the venerable Swordfish and Albacore but because of development problems did not begin carrier operations until mid-1944. The Barracuda gained some fame for sinking battleship *Tirpitz* at her moorings in Norway. The Mark II, equipped with radar, also served from British carriers in ASW operations. (Simon Parry)

This Grumman Martlet IV is seen aboard a British carrier with a British destroyer-escort in the background. The FAA acquired over 1,400 Martlets. (Simon Parry)

To replace its Fairey Albacore and Barracuda, the British ordered over 400 Grumman Avengers for service on their fleet carriers and ASW "jeep" carriers. The FAA renamed it the Tarpon. Two of the reliable aircraft are shown here (above) in flight near the Grumman Long Island Plant during an acceptance flight and (below) during an engine test run. (NARS)

CHAPTER 6

A GRIM FINAL ACCOUNTING

In Samuel Eliot Morison's **History of United States Naval Operations in World War II, Vol. X**, Atlantic-Little Brown, 1947, he estimates that German (and a few Italian) submarines sank 2,828 merchant vessels and 187 Allied warships in five years of war. For this the German Navy paid a commensurately high price. Clay Blair, in his most comprehensive two volume work, **Hitler's U-Boat War**, Random House, 1998, declares that 648 U-boats were lost at sea while on patrol, sunk by surface escorts, aircraft, or a combination of both. It is estimated that over 27,400 U-boat crewmen were lost and another 5,000 taken prisoners.

Losses among ASW units were high. They operated alone, over water, far from land. The two Army Air Force units operating from St. Eval, England, and French Morocco lost twenty-one aircraft to the following causes according to Max Schoenfeld, **Stalking the U-boat,** Smithsonian Press, 1995:

Enemy aircraft engagements	7
Unknown, combat operations	3
U-boat AA	1
Crashes/accidents	10

148 airmen perished in these aircraft losses.

Unit records for Fleet Air Wing 7 squadrons flying from England, VB-103, VB-105 and VB-110, show thirty-four losses as follows:

Enemy aircraft engagements	4
Unknown, combat operations	6
Crashes/accidents	22
Drone operations*	2

191 crewmen perished in these aircraft losses.

While these statistics are not necessarily representative of all ASW units, they are indicative of the broad range of perils.

*** VB-110 participated in remote control drone bomber experiments where two aircraft were employed and lost. One was piloted by Lt. Joe Kennedy.**

By 1944 the combination of Allied coordination and technical advances had doomed Hitler's U-boat offensive. Above: U-243 was first sighted ninety miles west of Brest, France on 8 Jul '44 by a Sunderland of 10 Sq RAAF, F/O Willis B. Tilley commanding. The U-boat put up a fight but Tilley's guns silenced the AA as he maneuvered low for a six depth charge straddle. Below: As the sub begins to settle by the stern another Sunderland and a PB4Y of VB-105 (Lt. Aurelian H. Cooledge) arrived on the scene and disgorged more bombs and machine gun rounds on the stricken vessel. The hunters then dropped life rafts and supplies for the thirty-eight survivors. Twelve men went down with U-234. (Don Schierenbeck)

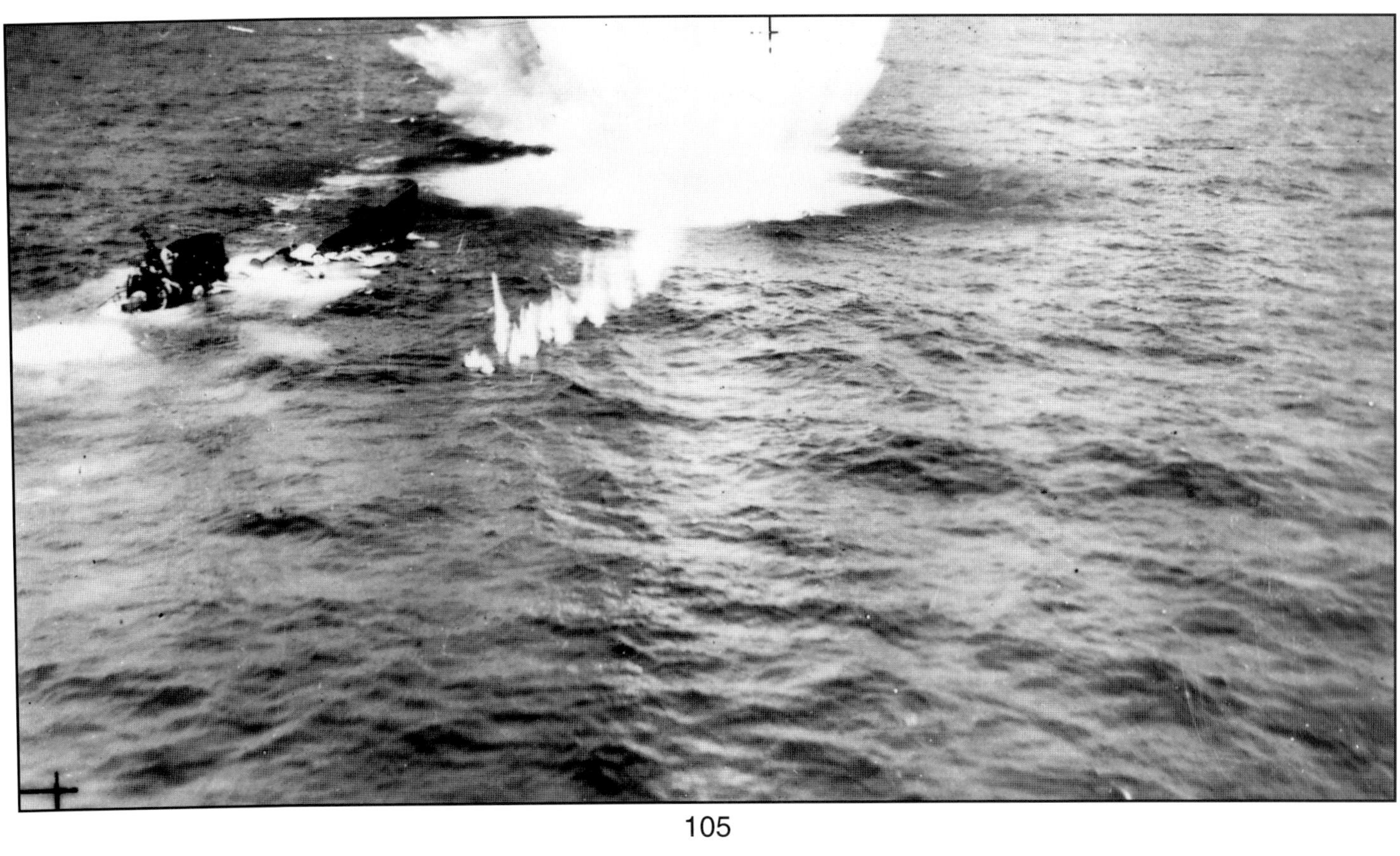

On the night of 16 May '44 a Catalina of 333 Sq (Norway) encountered U-668 off the coast of Norway. As Sub-Lt. Hartmann's aircraft FP121 dove to the attack, machine guns blazing, the sub fired back with twin-20s and a 37mm cannon. As the aircraft approached, the U-boat skipper ordered hard right rudder and the depth charges failed to cause any damage. Not so the Catalina. Cannon shells punctured the hull, damaged direction finding apparatus, and killed a gunner. The PBY returned to base in Scotland where it settled on its hull in the river Tay after landing. (Lorang Sjonnesen)

Short Sunderlands replaced some Norwegian Catalinas. WH-F of 330 Sq is seen here at rest in the Shetland Islands at Sollum Voe. Note the radar housing bubble under the left wing. (Egil Johansen)

Above: Attacked by Sunderland of RCAF 422 Sq near a convoy west of Ireland on 10 Mar '44, U-625 was straddled by depth charges as it maneuvered and fired its AA guns. The large splashes, upper right, mark the fall of bombs. The sub dove briefly but then resurfaced, mortally wounded. Below: The Sunderland took this post action photo of U-625, dead in the water, sinking by the stern with the crew abandoning ship. By the time surface ships reached the location however, all fifty-three German submariners had vanished. Commander on this mission was RAF check pilot Sidney Butler. (Canadian Armed Forces)

De Havilland Mosquitos prowling the North Sea attacked surface shipping and U-boats when the opportunity was presented. Above: A Mark VI Mosquito of RAF 143 Sq flown by its CO, Wg. Cdr. Foxley Norris, mounts four machine guns and four 20mm cannons in the nose plus eight 60 lb. under-wing rockets. Below: On 9 Apr 1945 Mosquitoes of Banff Strike Wing's 143, 235, and 248 Squadrons were patrolling the Kattegut Channel when they spotted the wakes of shallow running U-804 and U-1065 outbound from Kiel. A short distance away was U-843 inbound to Kiel from the Dutch East Indies. With cannon and rockets the Mosquitoes sank all three subs and the entire complement of 144 German crew members. The explosions that doomed the subs also downed one attacker and damaged three others. (Andrew Bird)

RAF 210 Sq was one of the few Coastal Command units that flew Catalinas from 1941 to 1945. Above: An early model with all her ASV antenna carries a load of ordnance under each wing. Below: The proud crew of DA-X. Led by Flt. Lt. K.M. Murray, they scored the last direct action U-boat sinking of the war. Their victim was U-320 which was so badly damaged by Murray's depth charge attack on 7 May '45 that it was scuttled off Bergen, Norway. 210 Sq sank eight U-boats. (David Legg)

The venerable Catalina soldiered on through the campaign.This 1945 photo shows a Canadian Canso and its crew at Dartmouth, near Halifax, Nova Scotia. The ordnance under the port wing is an acoustic torpedo called "Proctor" by the RAF and RCAF and "Fido" by the U.S. Navy. A pair of depth charges can be seen under the starboard wing. (Canadian Armed Forces)

When Adm. Karl Doenitz signaled his U-boat fleet that the war had ended, he ordered them to proceed on the surface toward the nearest Allied port or naval units. U-239 is observed here, a black surrender flag on her mast, dutifully carrying out their final order. It was 9 May '45 as seen by a Liberator of VPB-112, Lt. Fred Schaum, commanding. (Don Schierenbeck)

One of the surviving victors, a much traveled Mark III Short Sunderland flying boat of RAF 202 Sq shown here on its beaching gear at Belfast during a VE-day celebration. (Ulster Avn. Society)

AIR COMBAT PHOTO HISTORY SERIES

Volume I	Low Level Attack: The Pacific
Volume II	Bombs, Torpedoes and Kamikazes
Volume III	Low Level Attack: The Mediterranean & Europe
Volume IV	Atlantic Air War: Sub Hunters vs U-boats